My class sent a letter to a candy shop. We would love to see the people make candy.

We went to the candy shop right after school. Only one other class has come here.

The candy shop is run by two
people. They are a mother and her
brother. They have a number of
pretty shops under their name.

One candy is light in color. Another
candy comes in many bright colors.

There is crispy candy and soft candy. There is pretty little penny candy too.

Candy is stacked high up on shelf after shelf. If the shop had any more candy, we could not move! That would be a funny sight!

The candy shop sells candy to
other shops. They fly some candy
by plane.

Try the taffy. This candy shop is
better than any other!